HOUGHTON MIFFLIN HARCOURT

Go Math!

Intensive Intervention

 **RtI** Response to Intervention Tier 3

Skill Packs

Grade 2

Houghton
Mifflin
Harcourt

www.hmhschool.com

Contents

Skills

Learn the Math

You can join groups to find how many there are in all.

4 apples   and   2 apples

Count all the apples. 4 and 2 is __6__ .

_____ peaches   and   _____ peach

Count all the peaches. 3 and 1 is _____ .

Write how many there are in all.

1.

2 and 3 is ___5___.

2.

4 and 1 is _____.

3.

1 and 1 is _____.

4.

3 and 1 is _____.

5.

2 and 1 is _____.

Name_____

Learn the Math

You can write an addition sentence with + and = .

Vocabulary

plus (+)

is equal to (=)

sum

addition sentence

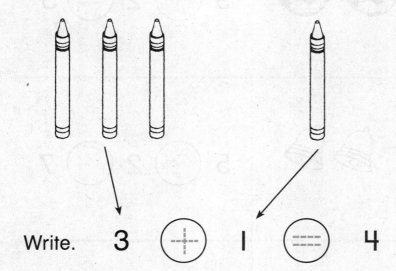

Write.   3 ⊕ 1 ⊜ 4

Say.   3 plus 1 is equal to 4.

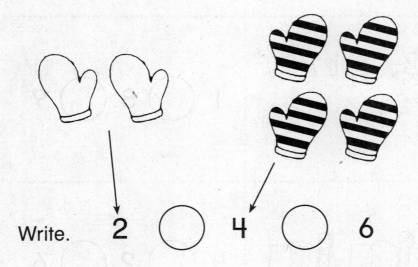

Write.   2 ◯ 4 ◯ 6

Say.   2 plus 4 is equal to 6.

Use the picture. Write the addition sentence.

1.

3 ⊕ 2 ⊜ 5

2.

5 ◯ 2 ◯ 7

3.

2 ◯ 2 ◯ 4

4.

 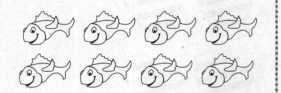

1 ◯ 8 ◯ 9

5.

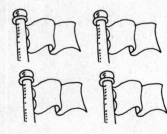 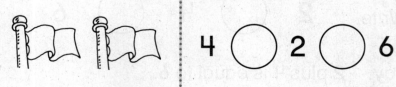

4 ◯ 2 ◯ 6

Name_____

Learn the Math

You can find how many are left.

_____4_____ rabbits   take away   _____1_____ rabbit

Count the rabbits that are left.

4 take away 1 is _____3_____ .

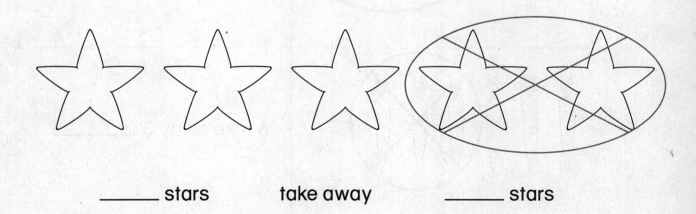

_____ stars   take away   _____ stars

Count the stars that are left.

5 take away 2 is _____ .

Write how many are left.

1.

6 take away 2 is ___4___.

2.

5 take away 1 is _____.

3.

4 take away 2 is _____.

4.

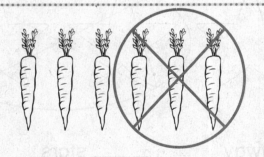

6 take away 3 is _____.

5.

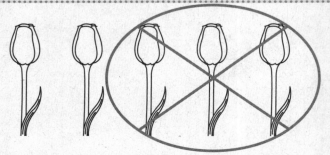

5 take away 3 is _____.

Learn the Math

You can write a subtraction sentence
with − and =.

Vocabulary

minus (−)

is equal to (=)

subtraction sentence

Write.   4   2   2

Say.   4 minus 2 is equal to 2.

Write.   5   1   4

Say.   5 minus 1 is equal to 4.

Complete the subtraction sentence.

1.

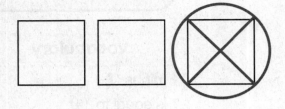   3 1 _2_

2.

   6 ◯ 3 ◯ ___

3.

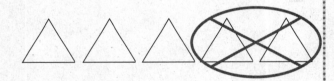   5 ◯ 2 ◯ ___

4.

   7 ◯ 4 ◯ ___

5.

   10 ◯ 5 ◯ ___

Name_____

Learn the Math

You can use counters to show numbers.

**How can you show 6 with counters?
Draw counters to show 6.**

Draw counters to show 9.

Draw counters to show 14.

Draw counters to show the number.

1.

8

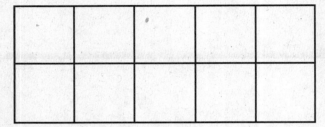

2.

12

 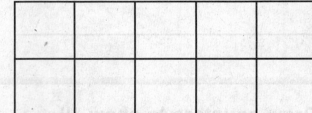

3.

15

4.

7

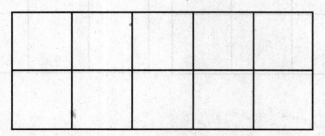

Name_____

Learn the Math

You can use groups of 10 to find how many.

How many  **are there?**

Count groups of 10 first.

How many groups of 10 🌙 are there? ___2___   2 tens = 20

How many other 🌙 are there? ___4___   4 ones = 4

How many 🌙 are there in all? ___24___   2 tens 4 ones = 24

How many ERASER **are there?**

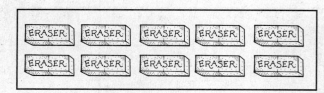

Count the group of 10 first.

Then count the rest.   I ten = ____

There are _____ ERASER in all.   5 ones = ____

                                                  I ten 5 ones = ____

© Houghton Mifflin Harcourt

Write how many.

1.

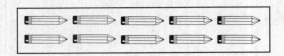   15

2.

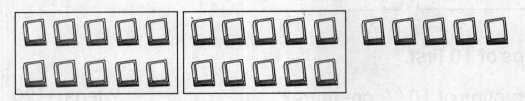   ____

3.

    ____

4.

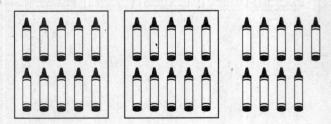   ____

5.

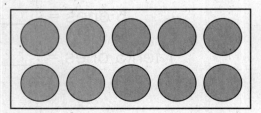    ____

Name_____

Learn the Math

Find the greater number.

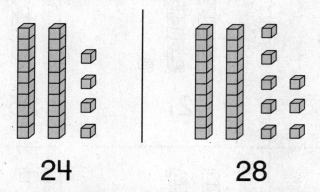

24   28

Step 1 Compare the tens.

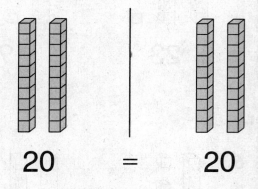

20   =   20

Step 2 If the tens are the same, compare the ones.

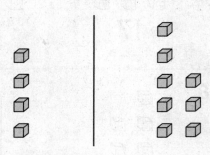

4   is less than   8

So, __28__ is the greater number.

Compare. Circle the greater number.

1.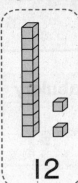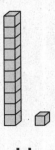

12   11

2.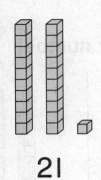

21   25

3.

7   14

4.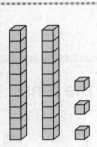

23   19

Compare. Circle the lesser number.

5.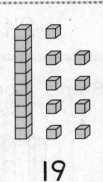

13   5

6.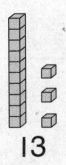

17   27

7.

16   21

8.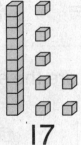

9   15

Name_____

Learn the Math

You can use a number line to find the number that is before, after, or between other numbers.

Vocabulary

before
between
after
number line

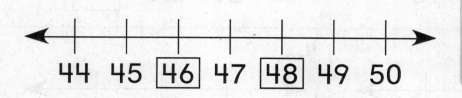

44 45 46 47 48 49 50

What number is just before 46?
Find 46.

45 is just before 46.

What number is just after 48?
Find 48.

49 is just after 48.

What number is between 46 and 48?
Find 46 and 48.

_____ is between 46 and 48.

What number is just after 24?

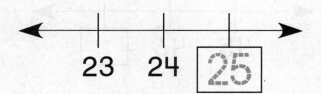

23 24 25

_____ is just after 24.

© Houghton Mifflin Harcourt

Write the number that is just before, just after, or between.

1.

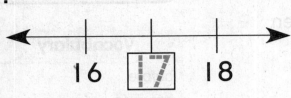

16   17   18

2.

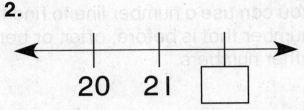

20   21   ☐

3.

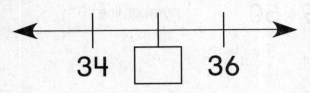

34   ☐   36

4.

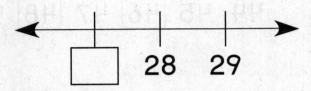

☐   28   29

5.

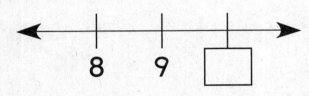

8   9   ☐

6.

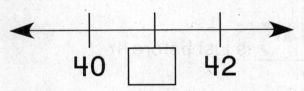

40   ☐   42

7.

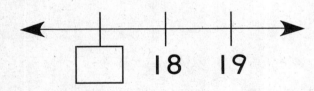

☐   18   19

8.

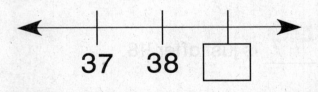

37   38   ☐

9.

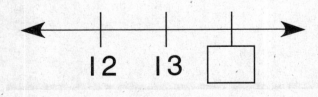

12   13   ☐

10.

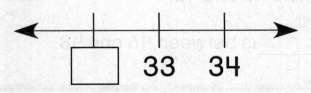

☐   33   34

11.

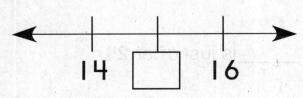

14   ☐   16

12.

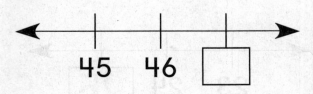

45   46   ☐

Name_____

Learn the Math

Add 1. Look for a pattern.

Vocabulary

pattern

  $1 + 1 = \underline{2}$

   $2 + 1 = \underline{3}$

  $3 + 1 = \underline{4}$

Add 2. Look for a pattern.

 $1 + 2 = \underline{3}$

            $2 + 2 = \underline{}$

 $3 + 2 = \underline{}$

Count how many. Draw one more.
Write how many in all.

1.

        3   +   1   =   __4__

2.

        4   +   1   =   _____

3.

        5   +   1   =   _____

4.

        6   +   1   =   _____

Add 2. Complete each addition sentence.

5. 1 + __2__ = _____   6. 2 + ____ = _____

7. 3 + ____ = _____   8. 4 + ____ = _____

9. 5 + ____ = _____   10. 6 + ____ = _____

Learn the Math

Subtract 1. Look for a pattern.

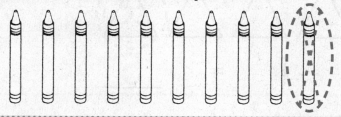

$$10 - 1 = \underline{9}$$

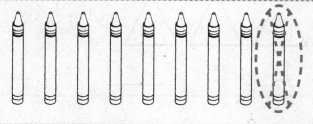

$$9 - 1 = \underline{8}$$

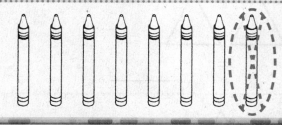

$$8 - 1 = \underline{7}$$

Subtract 2. Look for a pattern.

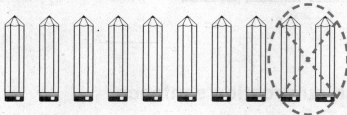

$$10 - 2 = \underline{8}$$

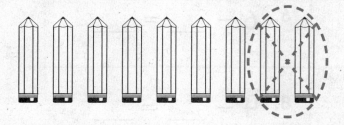

$$9 - 2 = \underline{}$$

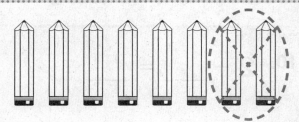

$$8 - 2 = \underline{}$$

Circle and mark an X to subtract.
Complete the subtraction sentence.

1.

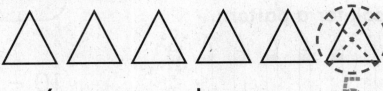

6   –   1   =   5

2.

5   –   1   =   ___

3.

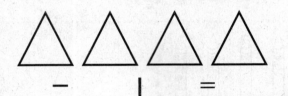

4   –   1   =   ___

4.

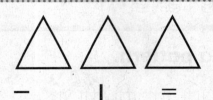

3   –   1   =   ___

Subtract 2. Complete each subtraction sentence.

5. 7 – 2 = ___

6. 6 – ___ = ___

7. 5 – ___ = ___

8. 4 – ___ = ___

9. 3 – ___ = ___

10. 2 – ___ = ___

Name_____

Learn the Math

You can use a number line to count on.

$2 + 6 = \underline{?}$

Which addend is greater? _6_

So, start on 6.

Then move 2 spaces to the right.

Vocabulary

count on
addend

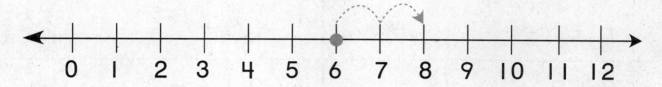

So, $2 + 6 = \underline{8}$.

Use the number line to count on.

$4 + 3 = \underline{?}$

Start on _4_, the greater addend.

Move 3 spaces to the right.

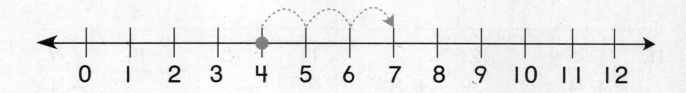

So, $4 + 3 = \underline{}$.

Do the Math

Use the number line to count on.
Write the sum.

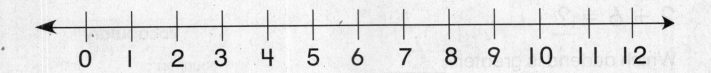

1. $9 + 2 = $ <u>11</u>

2. $3 + 1 = $ ___

3. $8 + 3 = $ ___

4. $3 + 6 = $ ___

5. $2 + 1 = $ ___

6. $7 + 2 = $ ___

7.
$$\begin{array}{r} 7 \\ +3 \\ \hline \end{array}$$

8.
$$\begin{array}{r} 6 \\ +1 \\ \hline \end{array}$$

9.
$$\begin{array}{r} 1 \\ +8 \\ \hline \end{array}$$

10.
$$\begin{array}{r} 9 \\ +3 \\ \hline \end{array}$$

11.
$$\begin{array}{r} 3 \\ +4 \\ \hline \end{array}$$

12.
$$\begin{array}{r} 2 \\ +8 \\ \hline \end{array}$$

13.
$$\begin{array}{r} 6 \\ +2 \\ \hline \end{array}$$

14.
$$\begin{array}{r} 9 \\ +1 \\ \hline \end{array}$$

Name_____

Learn the Math

You can add doubles.
A doubles fact has two addends
that are the same.

Vocabulary

doubles fact
addend

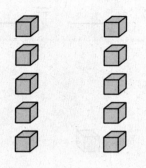

$$\underline{5} + \underline{5} = \underline{10}$$

Write the doubles fact.

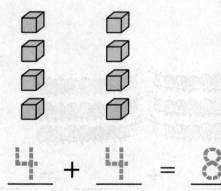

$$\underline{4} + \underline{4} = \underline{8}$$

$$\underline{} + \underline{} = \underline{}$$

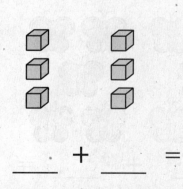

$$\underline{} + \underline{} = \underline{}$$

$$\underline{} + \underline{} = \underline{}$$

Write the doubles fact.

1.

$$\underline{3} + \underline{3} = \underline{6}$$

2.

$$\underline{} + \underline{} = \underline{}$$

3.

$$\underline{} + \underline{} = \underline{}$$

4.

$$\underline{} + \underline{} = \underline{}$$

Write the doubles fact for each picture.

5.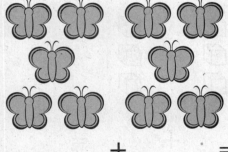

$$\underline{} + \underline{} = \underline{}$$

6.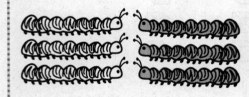

$$\underline{} + \underline{} = \underline{}$$

7.

$$\underline{} + \underline{} = \underline{}$$

8.

$$\underline{} + \underline{} = \underline{}$$

Name_____

Learn the Math

You can use a doubles-plus-one fact
to find a sum.

$3 + 3$

The addends are the same.

It is a doubles fact.

$3 \quad + \quad 3 \quad = \quad \underline{6}$

$3 + 4$

4 is one more than 3.

It is a doubles-plus-one fact.

Think: 3 + 3 and 1 more.

$3 \quad + \quad 4 \quad = \quad \underline{7}$

Find the sum.

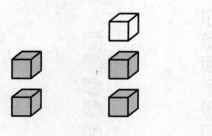

$2 + 3$

3 is one more than 2.

It is a doubles-plus-one fact.

Think: 2 + 2 and 1 more.

$2 \quad + \quad 3 \quad = \quad \underline{}$

Use 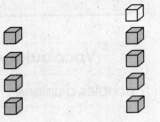 . Write the addition sentence.

1.

$$\underline{4} + \underline{5} = \underline{9}$$

2.

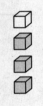

$$\underline{} + \underline{} = \underline{}$$

3.

$$\underline{} + \underline{} = \underline{}$$

4.

$$\underline{} + \underline{} = \underline{}$$

5.

$$\underline{} + \underline{} = \underline{}$$

6.

$$\underline{} + \underline{} = \underline{}$$

7.

$$\underline{} + \underline{} = \underline{}$$

8.

$$\underline{} + \underline{} = \underline{}$$

Name_____

Learn the Math

You can add in any order.
The sum is the same.

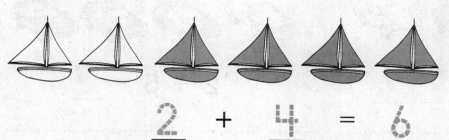

$$2 + 4 = 6$$

addend　　addend　　sum

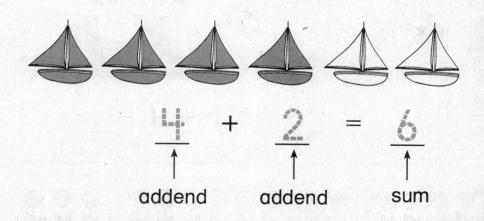

$$4 + 2 = 6$$

addend　　addend　　sum

Write the sum.

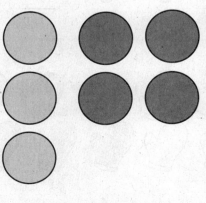

$$3 + 4 = \underline{}$$

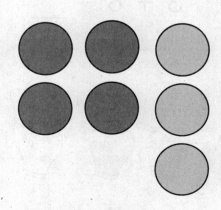

$$4 + 3 = \underline{}$$

Write the sum.

1.

$5 + 2 = \underline{7}$ $2 + 5 = \underline{7}$

2.

   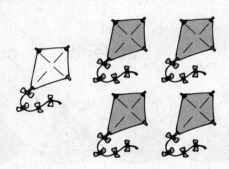

$4 + 1 = \underline{}$ $1 + 4 = \underline{}$

3.

$3 + 6 = \underline{}$ $6 + 3 = \underline{}$

4.

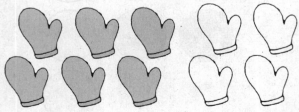   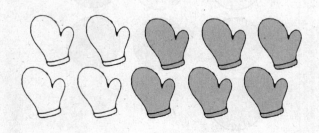

$6 + 4 = \underline{}$ $4 + 6 = \underline{}$

Learn the Math

You can use base-ten blocks to add tens.

40 + 20 = __?__

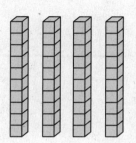   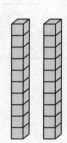

4 tens   +   2 tens   = __6__ tens

40   +   20   = __60__

---

30 + 50 = __?__

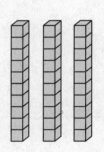   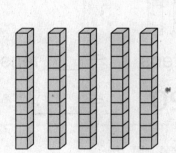

3 tens   +   5 tens   = ____ tens

30   +   50   = ____

© Houghton Mifflin Harcourt

Use base-ten blocks. Find each sum.

1.

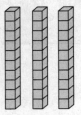 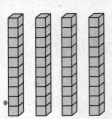

3 tens   +   4 tens   = __7__ tens

30   +   40   = __70__

2.

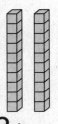 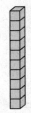

2 tens   +   1 ten   = ___ tens

20   +   10   = ___

3.

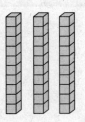 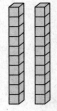

3 tens   +   2 tens   = ___ tens

30   +   20   = ___

4.

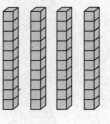 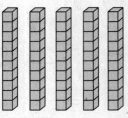

4 tens   +   5 tens   = ___ tens

40   +   50   = ___

Name_____

Learn the Math

You can count back on a number line
to subtract.

$$9 - 2 = \underline{?}$$

Step 1 Put your finger on 9.

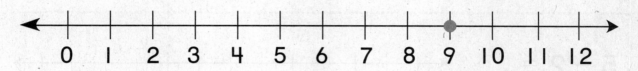

Step 2 Move 2 spaces to the left.

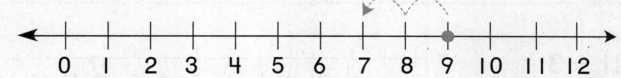

On what number do you stop? $\underline{7}$

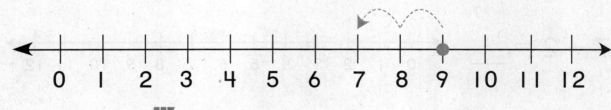

So, $9 - 2 = \underline{7}$.

Use the number line. Count back to subtract.

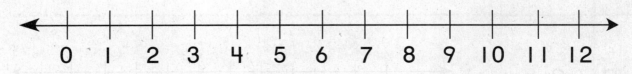

$$7 - 3 = \underline{}$$

Use the number line. Count back to subtract.

1. $6 - 2 =$ __4__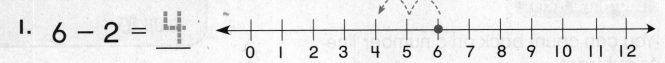

2. $10 - 1 =$ ___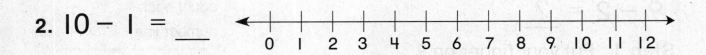

3. $5 - 2 =$ ___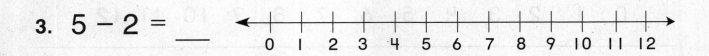

4. $11 - 3 =$ ___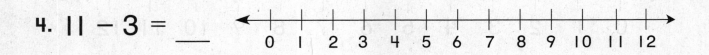

5. $7 - 2 =$ ___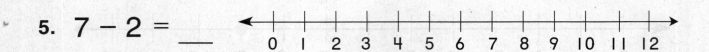

6. $8 - 1 =$ ___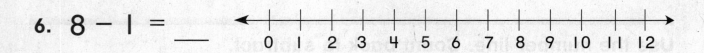

7. $9 - 3 =$ ___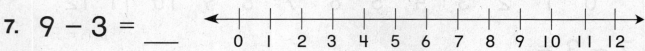

Name_____

Learn the Math

Related subtraction facts use the same three numbers.

Vocabulary

related subtraction facts

$$\begin{array}{r} 10 \\ -\ 3 \\ \hline 7 \end{array}$$

What numbers are used? _3_ , _7_ , _10_

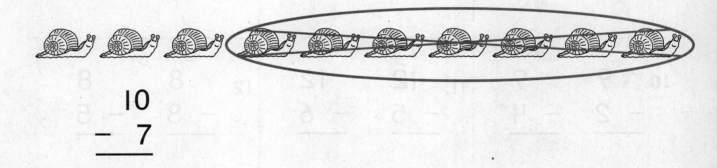

$$\begin{array}{r} 10 \\ -\ 7 \\ \hline \end{array}$$

What numbers are used? ___ , ___ , ___

So, $10 - 3 = 7$ and $10 - 7 = 3$ are related subtraction facts.

Subtract. Circle the pairs of related subtraction facts.

1.
$$\begin{array}{r} 9 \\ -\ 6 \\ \hline 3 \end{array} \quad \begin{array}{r} 9 \\ -\ 3 \\ \hline 6 \end{array}$$

2.
$$\begin{array}{r} 8 \\ -\ 2 \\ \hline \end{array} \quad \begin{array}{r} 8 \\ -\ 6 \\ \hline \end{array}$$

3.
$$\begin{array}{r} 10 \\ -\ 4 \\ \hline \end{array} \quad \begin{array}{r} 10 \\ -\ 5 \\ \hline \end{array}$$

4.
$$\begin{array}{r} 12 \\ -\ 7 \\ \hline \end{array} \quad \begin{array}{r} 12 \\ -\ 5 \\ \hline \end{array}$$

5.
$$\begin{array}{r} 8 \\ -\ 3 \\ \hline \end{array} \quad \begin{array}{r} 8 \\ -\ 4 \\ \hline \end{array}$$

6.
$$\begin{array}{r} 11 \\ -\ 6 \\ \hline \end{array} \quad \begin{array}{r} 11 \\ -\ 5 \\ \hline \end{array}$$

7.
$$\begin{array}{r} 12 \\ -\ 9 \\ \hline \end{array} \quad \begin{array}{r} 12 \\ -\ 3 \\ \hline \end{array}$$

8.
$$\begin{array}{r} 11 \\ -\ 3 \\ \hline \end{array} \quad \begin{array}{r} 11 \\ -\ 8 \\ \hline \end{array}$$

9.
$$\begin{array}{r} 10 \\ -\ 9 \\ \hline \end{array} \quad \begin{array}{r} 10 \\ -\ 1 \\ \hline \end{array}$$

10.
$$\begin{array}{r} 9 \\ -\ 2 \\ \hline \end{array} \quad \begin{array}{r} 9 \\ -\ 4 \\ \hline \end{array}$$

11.
$$\begin{array}{r} 12 \\ -\ 5 \\ \hline \end{array} \quad \begin{array}{r} 12 \\ -\ 6 \\ \hline \end{array}$$

12.
$$\begin{array}{r} 8 \\ -\ 3 \\ \hline \end{array} \quad \begin{array}{r} 8 \\ -\ 5 \\ \hline \end{array}$$

13.
$$\begin{array}{r} 12 \\ -\ 2 \\ \hline \end{array} \quad \begin{array}{r} 12 \\ -\ 8 \\ \hline \end{array}$$

14.
$$\begin{array}{r} 10 \\ -\ 8 \\ \hline \end{array} \quad \begin{array}{r} 10 \\ -\ 2 \\ \hline \end{array}$$

15.
$$\begin{array}{r} 9 \\ -\ 7 \\ \hline \end{array} \quad \begin{array}{r} 9 \\ -\ 2 \\ \hline \end{array}$$

Learn the Math

A fact family is a group of related addition and subtraction sentences. They all use the same numbers.

What is the fact family for 2, 3, and 5?

Step 1 Write related addition sentences.

$$\underline{2} + \underline{3} = \underline{5} \qquad \underline{3} + \underline{2} = \underline{5}$$

Step 2 Write related subtraction sentences.

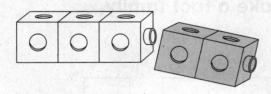 

$$\underline{5} - \underline{2} = \underline{3} \qquad \underline{5} - \underline{3} = \underline{2}$$

So, the fact family for 2, 3, and 5 is:

$$\underline{} + \underline{} = \underline{} \qquad \underline{} - \underline{} = \underline{}$$

$$\underline{} + \underline{} = \underline{} \qquad \underline{} - \underline{} = \underline{}$$

Add or subtract to complete the fact family.
Write the numbers in the fact family.

1. $6 + 3 = \underline{9}$ $9 - 6 = \underline{3}$

    $3 + 6 = \underline{9}$ $9 - 3 = \underline{6}$  3, 6, 9

2. $5 + 2 = \underline{}$ $7 - 5 = \underline{}$

    $2 + 5 = \underline{}$ $7 - 2 = \underline{}$ ☐, ☐, ☐

3. $1 + 4 = \underline{}$ $5 - 1 = \underline{}$

    $4 + 1 = \underline{}$ $5 - 4 = \underline{}$ ☐, ☐, ☐

4. $5 + 5 = \underline{}$ $10 - 5 = \underline{}$ ☐, ☐

Write the number sentences to make a fact family.

5. 4, 5, 9

 $4 + 5 = 9$ ☐ − ☐ = ☐

   ☐ + ☐ = ☐   ☐ − ☐ = ☐

6. 3, 7, 10

   ☐ + ☐ = ☐   ☐ − ☐ = ☐

   ☐ + ☐ = ☐   ☐ − ☐ = ☐

Name_____

Learn the Math

You can use addition facts to help you subtract.

8 − 6 = ?

Think: 6 + 2 = 8

Since 6 + ___ = 8, then 8 − 6 = ___.

---

10 − 3 = ?

Think: 3 + ___ = 10

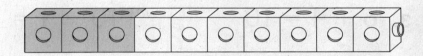

Since 3 + ___ = 10, then 10 − 3 = ___.

---

9 − 6 = ?

Think: 6 + ___ = 9

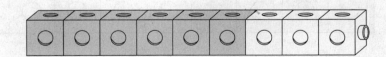

Since 6 + ___ = 9, then 9 − 6 = ___.

Find each difference. Use the addition fact to help you.

1. $3 + 4 = 7$ $\qquad$ $7 - 4 = \underline{3}$

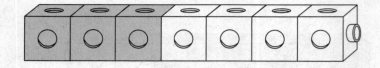

2. $5 + 3 = 8$ $\qquad$ $8 - 3 = \underline{\hphantom{00}}$

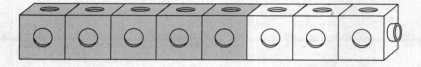

3. $6 + 4 = 10$ $\qquad$ $10 - 4 = \underline{\hphantom{00}}$

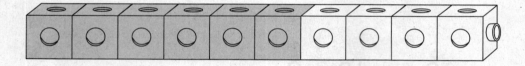

4. $4 + 5 = 9$ $\qquad$ $9 - 5 = \underline{\hphantom{00}}$

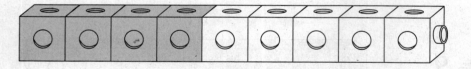

5. $7 + 5 = 12$ $\qquad$ $12 - 5 = \underline{\hphantom{00}}$

Name_____

Learn the Math

You can use base-ten blocks to subtract tens.

50 − 20 = _?_

How many tens are there in all? _5_ tens

How many tens are taken away? _2_ tens

How many tens are left? _3_ tens

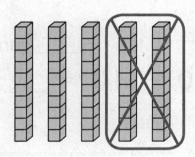

5 tens − 2 tens = _3_ tens

50 − 20 = _30_

---

60 − 40 = _?_

How many tens in are there all? ____ tens

How many tens are taken away? ____ tens

How many tens are left? ____ tens

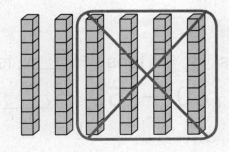

6 tens − 4 tens = ____ tens

60 − 40 = ____

Use base-ten blocks. Find each difference.

1.

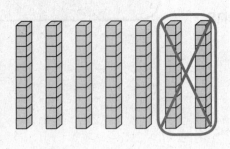

7 tens $-$ 2 tens $=$ __5__ tens

$70 - 20 = \underline{50}$

2.

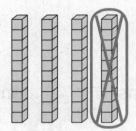

4 tens $-$ 1 ten $=$ ____ tens

$40 - 10 =$ ___

3.

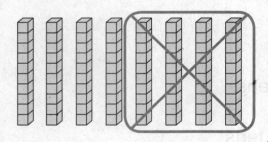

8 tens $-$ 4 tens $=$ ____ tens

$80 - 40 =$ ___

4.

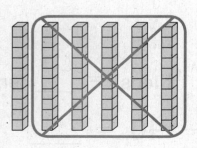

6 tens $-$ 5 tens $=$ ____ ten

$60 - 50 =$ ___

Name_____

Learn the Math

Erin has some puppets.

She makes a tally table about her puppets.

Erin's Puppets	
Puppet	**Tally**
🐯 tiger	|||
🐻 bear	||||/
🐵 monkey	||

Read the title.

What is the tally table about?

Erin's Puppets

Read the label for each row.

What kinds of puppets does Erin have?

tigers, bears, and monkeys

Read the tally marks.

How many of each puppet does Erin have?

 3   ____   ____

Compare.

Of which kind of puppet is there the most? _____

Do the Math

Use the tally table.

Vegetables in Maya's Garden	
Vegetable	**Tally**
carrot	|||| |||
tomato	||||	
pepper	|||

1. What is the tally table about?

Vegetables in Maya's Garden

- -

2. How many of each kind are in Maya's garden?

 _____   _____   _____

- -

3. Which of these is there the most?

- -

4. Which of these is there the least?

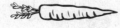

- -

5. How many vegetables does Maya have in all?

_____ vegetables

Name_____

Learn the Math

You can read a picture graph. Use the picture graph to answer the question.

Our Favorite Flowers

🌹 rose	🌹	🌹	🌹	🌹	🌹	🌹				
🌼 daisy	🌼	🌼	🌼	🌼						
🌷 tulip	🌷	🌷	🌷	🌷	🌷					

How many children chose ?

Step 1

Read the question.   How many children chose ?

Step 2

Count how many .

I, __2__, __3__, __4__

Step 3

Write how many.   _____ children

So, _____ children chose .

© Houghton Mifflin Harcourt

Use the picture graph.

Our Favorite Snacks								
apple	apple	apple	apple					
GRANOLA	GRANOLA	GRANOLA	GRANOLA	GRANOLA	GRANOLA	GRANOLA	GRANOLA	
YOGURT	YOGURT	YOGURT	YOGURT	YOGURT	YOGURT			

1. How many children chose ? ___3___ children

2. How many children chose 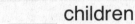 ? _____ children

3. How many children chose ? _____ children

4. Circle the snack that most children chose.

Learn the Math

Evie has a bag of tiles.

She has 4 gray tiles and 1 black tile.

Evie pulls a tile.

Which color is Evie more likely to pull?

Step 1

Count how many tiles there are of each color.

4   1

Step 2

Circle the color there is more of.

Step 3

Think: There are more gray tiles.
Circle your prediction.

So, Evie is more likely to pull a _____ tile.

Do the Math

Use the pictures to make a prediction.

1. Alex pulls a sock from the bag. Which color sock is Alex more likely to pull?

   Count the socks.

   Alex is more likely to pull  .

---

2. Devin pulls a marble from the bag. Which color marble is Devin more likely to pull?

   Count the marbles.

   Devin is more likely to pull   .

---

3. Delia pulls a ribbon from the bag. Which color ribbon is Delia more likely to pull?

   Count the ribbons.

   ____   ____

   Delia is more likely to pull   .

Name_____

Learn the Math

You can show the values of pennies, nickels, and dimes.

Show the value of one dime with pennies.

 =

10 cents

1 dime

__10__ pennies

Show the value of one dime with nickels.

 =

10 cents

1 dime

____ nickels

Show the value of one nickel with pennies.

 =

5 cents

1 nickel

____ pennies

Do the Math

Draw pennies to show equal value.

1.

 =

__5__ pennies

2.

 =

____ pennies

Draw nickels to show equal value.

3.   =

____ nickel

4. =

____ nickels

Name_____

Learn the Math

You can use a picture to skip-count.

How many grapes are there?
Skip-count by fives.

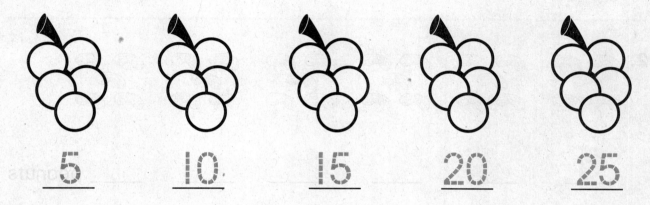

5   10   15   20   25

There are __25__ grapes.

How many cherries are there?
Skip-count by tens.

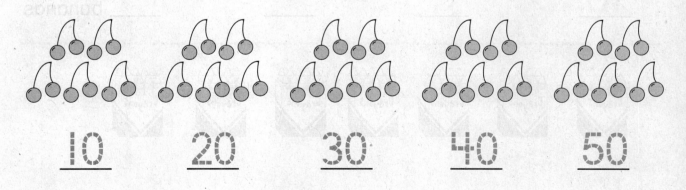

10   20   30   40   50

There are _____ cherries.

Skip-count by fives or tens. Write how many.

1.

   5   10   15   20   25   30 blueberries

2.

   5 ___ ___ ___ ___ ___ peanuts

3.

   10 ___ ___ ___ bananas

4.

   10 ___ ___ ___ ___ ___ ___ crayons

Name_____

Learn the Math

You can count to find the value of money.

Count by ones. Write the total value.

Vocabulary

penny   dime
nickel

total value

¢, ¢, 3 ¢

 3 ¢

Count by fives. Write the total value.

_____¢, _____¢, _____¢, _____¢

☐ ¢

Count by tens. Write the total value.

_____¢, _____¢, _____¢, _____¢, _____¢

☐ ¢

**Count by tens. Count by fives. Then count by ones.
Write the total value.**

_____¢, _____¢, _____¢, _____¢, _____¢, _____¢

☐ ¢

Count. Write the total value.

1.

   total value

__1__ ¢, __2__ ¢, __3__ ¢, __4__ ¢, __5__ ¢, __6__ ¢ | 6 | ¢

2.

____ ¢, ____ ¢, ____ ¢, ____ ¢, ____ ¢ | | ¢

3.

____ ¢, ____ ¢, ____ ¢ | | ¢

4.

____ ¢, ____ ¢, ____ ¢, ____ ¢, ____ ¢, ____ ¢ | | ¢

5.

____ ¢, ____ ¢, ____ ¢, ____ ¢, ____ ¢, ____ ¢ | | ¢

Name_____

Learn the Math

The value of one quarter is 25 cents.

   or

I quarter = 25 cents or 25¢

Count. Write the total value.
Start with the quarter. Count dimes by tens.
Count nickels by fives. Count pennies by ones.

   total value

$\underline{25}$ ¢, $\underline{35}$ ¢, $\underline{40}$ ¢, $\underline{45}$ ¢, $\underline{46}$ ¢ $\boxed{46}$ ¢

____ ¢, ____ ¢, ____ ¢, ____ ¢, ____ ¢, ____ ¢ $\boxed{}$ ¢

Count. Write the total value.

1.

total value

_25_ ¢, _35_ ¢, _45_ ¢, _55_ ¢, _56_ ¢, _57_ ¢ | 57 | ¢

2.

_____ ¢, _____ ¢, _____ ¢, _____ ¢, _____ ¢ | | ¢

3.

_____ ¢, _____ ¢, _____ ¢, _____ ¢, _____ ¢ | | ¢

4.

_____ ¢, _____ ¢, _____ ¢, _____ ¢, _____ ¢, _____ ¢ | | ¢

5.

_____ ¢, _____ ¢, _____ ¢, _____ ¢, _____ ¢ | | ¢

Name_____

Learn the Math

You can use a clock to tell time.

Vocabulary

clock
minute hand
hour hand

Step 1
Look at the minute hand.

— minute hand

It points to 12 .

Step 2
Look at the hour hand.

— hour hand

It points to __7__ .

Step 3
Write the time.

__7__ o'clock

Write the time.

__9__ o'clock

____ o'clock

____ o'clock

Do the Math

Write the time.

1.

2 o'clock

2.

___ o'clock

3.

___ o'clock

4.

___ o'clock

5.

___ o'clock

6.

___ o'clock

7.

___ o'clock

8.

___ o'clock

9.

___ o'clock

© Houghton Mifflin Harcourt

Name_____

Learn the Math

You can estimate about how long it takes to do something.

Vocabulary

minute	estimate
hour	

tie your shoes

about 1 minute

cook dinner

about 1 hour

About how long does it take?
Circle your answer.

exercise in gym class

about one minute

(about one hour)

brush your teeth

about two minutes

about two hours

About how long does it take? Circle your answer.

1. write 1 to 10

about 1 minute

about 1 hour

2. watch a play

about 2 minutes

about 2 hours

3. make a sandwich

about 5 minutes

about 5 hours

4. read a story

about 10 minutes

about 10 hours

5. sing a song

about 3 minutes

about 3 hours

6. take a piano lesson

about 1 minute

about 1 hour

Name_____

Learn the Math

You can sort objects by color.
All of these are white.

You can sort objects by shape.
All of these are circles.

You can sort objects by size.
All of these are the same size.

Mark an X on the object that does not belong.

Mark an X on the one that does not belong.

1.

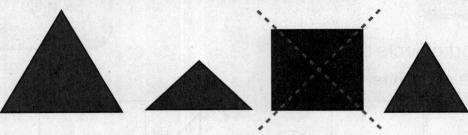

2.

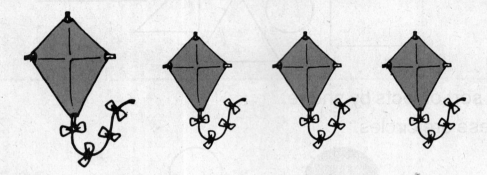

Circle the group in which the object belongs.

3.

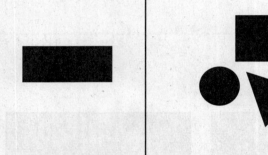   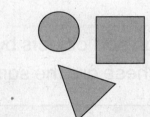

4.

Name_____

Learn the Math

These are three-dimensional shapes.

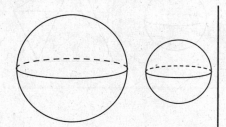

sphere

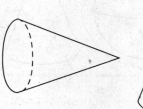

cone

Vocabulary

cone	cube
cylinder	
pyramid	
rectangular prism	
sphere	
three-dimensional	

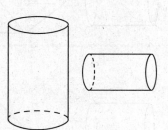

cylinder

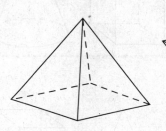

pyramid

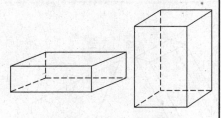

rectangular prism

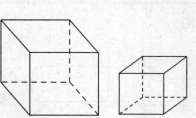

cube

Circle each pyramid.

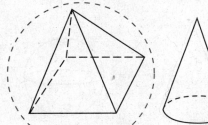

1. Circle each sphere.

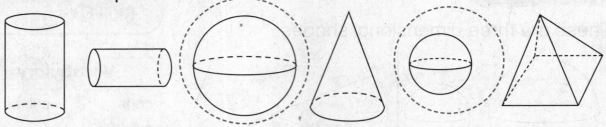

2. Circle each cylinder.

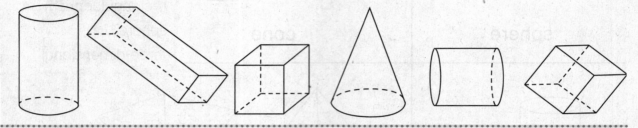

3. Circle each cube.

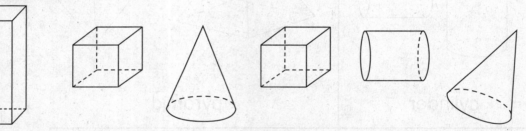

4. Circle each cone.

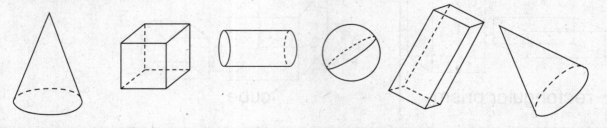

5. Circle each rectangular prism.

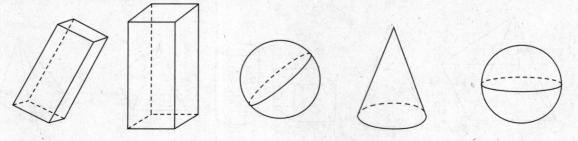

Name_____

Learn the Math

Some shapes have sides
and vertices.

Count the number of sides and vertices.

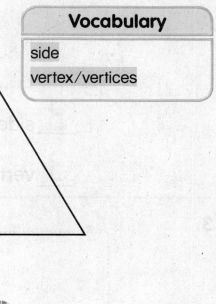

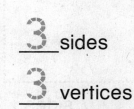

_4_ sides

_4_ vertices

_3_ sides

_3_ vertices

Trace each side. Circle each vertex.
Write how many sides and vertices.

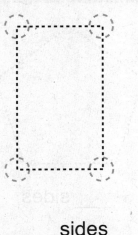

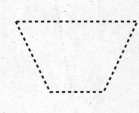

___ sides

___ vertices

___ sides

___ vertices

Trace each side. Circle each vertex. Write how many.

1.

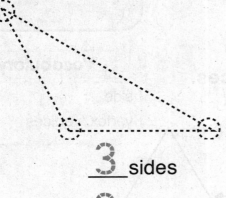

__3__ sides

__3__ vertices

2.

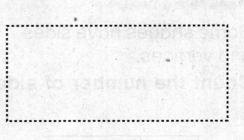

___ sides

___ vertices

3.

___ sides

___ vertices

4.

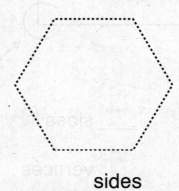

___ sides

___ vertices

5.

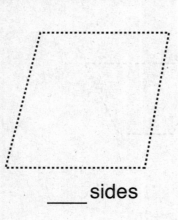

___ sides

___ vertices

6.

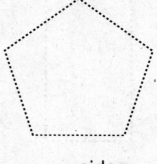

___ sides

___ vertices

Learn the Math

You can sort shapes by the number of sides
and vertices they have.

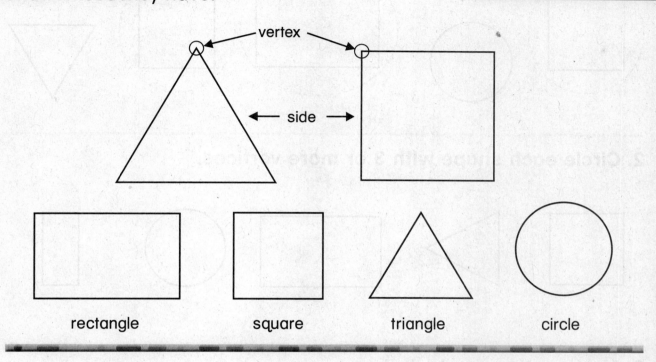

rectangle   square   triangle   circle

Circle each shape with 3 or more vertices.

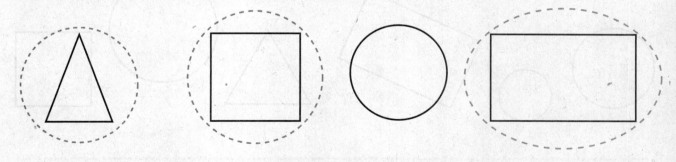

Mark an X on each triangle.

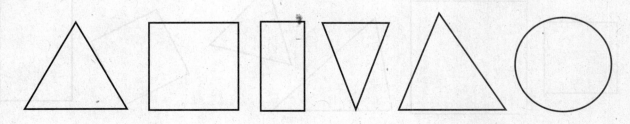

1. Circle each shape with 4 sides.

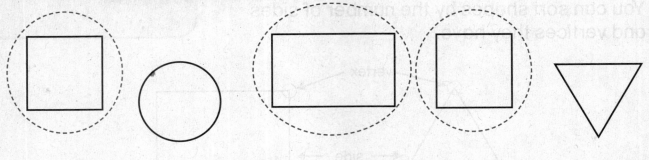

2. Circle each shape with 3 or more vertices.

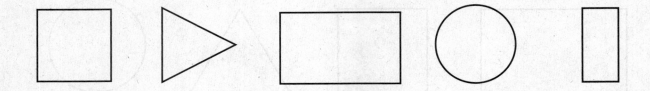

3. Mark an X on each circle.

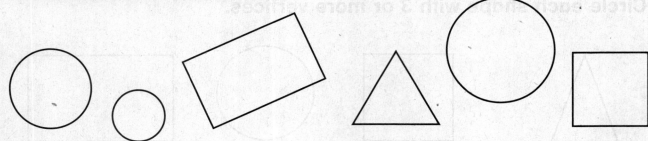

4. Mark an X on each square.

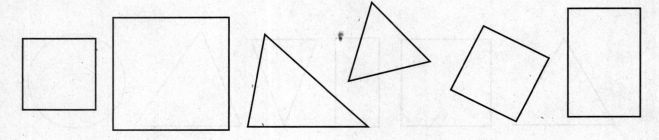

Name_____

Learn the Math

You can copy a pattern.

Step 1 Look for the pattern.

Step 2 Draw shapes to match.

Step 3 Color the shapes to match.

Color the shapes to copy the pattern.

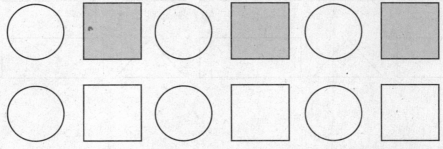

Color the shapes to copy the pattern.

1.

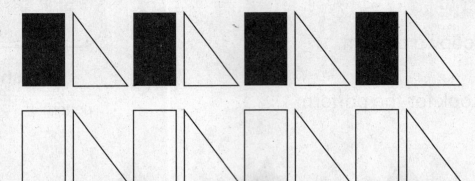

2.

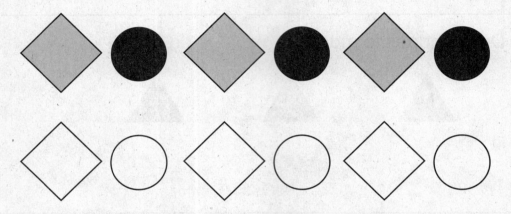

3.

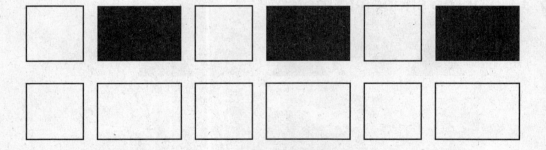

4.

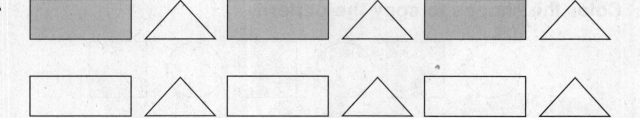

Name_____

Learn the Math

You can use a pattern unit to find what comes next in the pattern.

Vocabulary

pattern unit

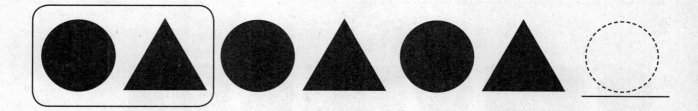

The pattern unit is circle, triangle.
Circle comes next.

Circle the pattern unit. Draw what comes next.

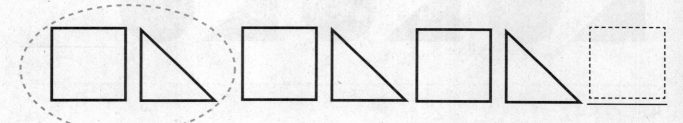

The pattern unit is square, triangle.
Square comes next.

The pattern unit is triangle, circle, square.
Triangle comes next.

Circle the pattern unit. Draw what comes next.

1.

2.

3.

4.

5.

Name_____

Learn the Math

You can compare lengths.

These objects are in order from shortest to longest.

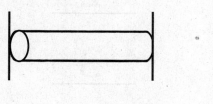 shortest

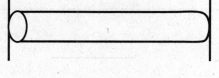

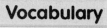 longest

Circle the objects that are in order from shortest to longest.

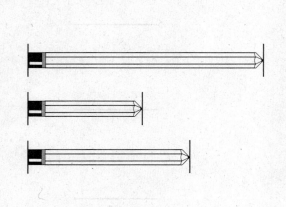

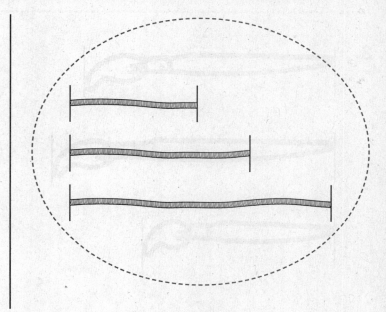

Order the objects from shortest to longest. Write 1, 2, and 3.

1.

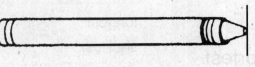

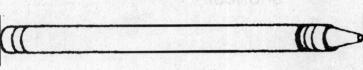

2.

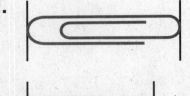

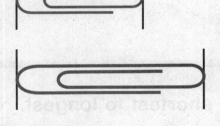

3.

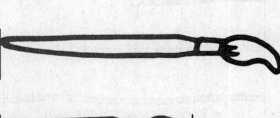

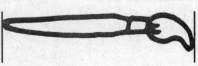

Learn the Math

Temperature is the measure of how hot or cold something is. A thermometer measures temperature.

Fahrenheit

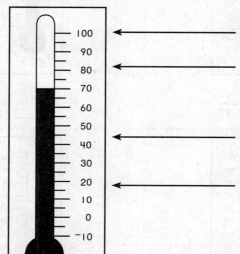

A very hot day is 100°F.

A warm day is 80°F.

A cool day is 45°F.

A very cold day is 18°F.

The temperature is _**70**_ °F.

Read each thermometer.
Write the temperature.

1.

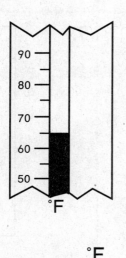

_____ °F

2.

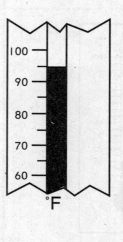

_____ °F

Do the Math

Read each thermometer.
Write the temperature.

1.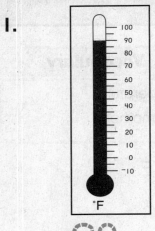

$\underline{90}$°F

2.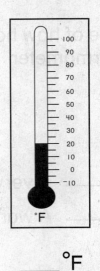

_____°F

3.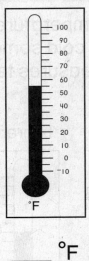

_____°F

4.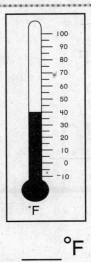

_____°F

5.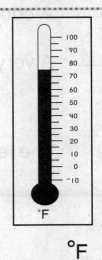

_____°F

6.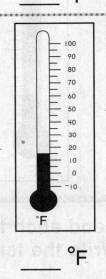

_____°F

7.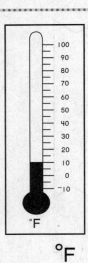

_____°F

8.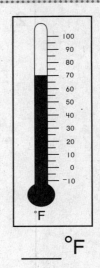

_____°F

9.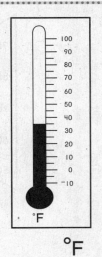

_____°F

Name_____

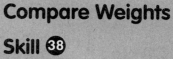

Learn the Math

You can hold objects to compare their weights.

Circle the object that is heavier.

Vocabulary

heavier
lighter

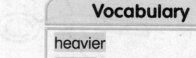

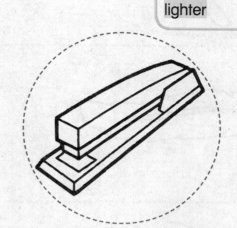

Hold one object in each hand.

Compare their weights.

The stapler is heavier.

Circle the object that is heavier.
Mark an X on the object that is lighter.

Circle the object that is heavier.
Mark an X on the object that is lighter.

1.

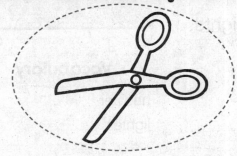 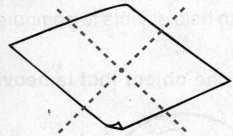

2.

3.

4.

 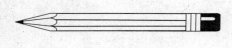

5.

Name_____

Learn the Math

You can compare containers by how much they hold.

Circle the container that holds more.

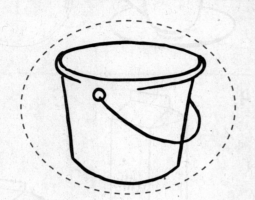

The  holds more than the  .

Circle the container that holds more.

Circle the container that holds less.

© Houghton Mifflin Harcourt

Circle the container that holds more.

1.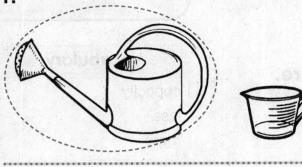

2.

3.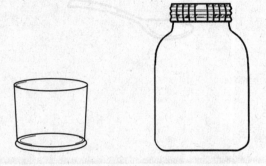

4.

Circle the container that holds less.

5.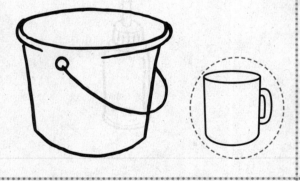

6.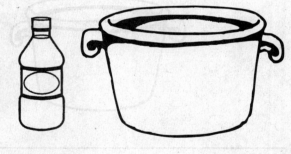

7.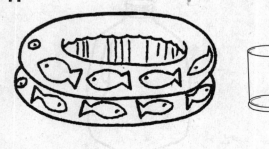

8.

Learn the Math

Equal parts are the same size.

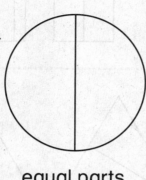 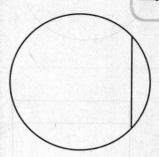

equal parts   unequal parts

Circle the shape that has 2 equal parts.

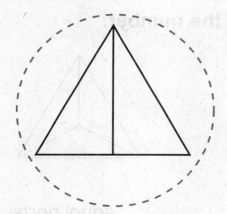

Count the equal parts.

  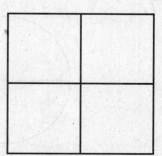

<u>2</u> equal parts   ____ equal parts   ____ equal parts

Circle the shape that has two equal parts.

1.

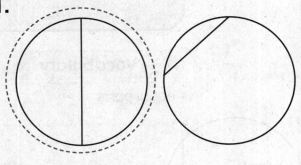

2.

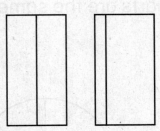

3.

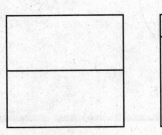

4.

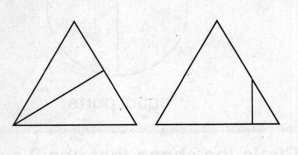

Count how many equal parts. Write the number.

5.

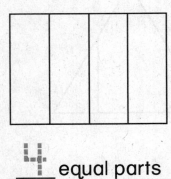

__4__ equal parts

6.

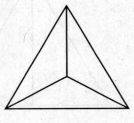

____ equal parts

7.

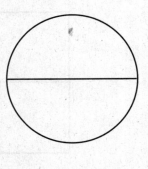

____ equal parts

8.

____ equal parts

Name_____

Learn the Math

Each of 2 equal parts of a whole is one half.

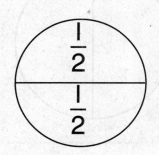

Vocabulary

one half $\frac{1}{2}$

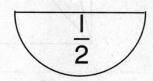

Two halves make one whole.

1 of 2 equal parts is $\frac{1}{2}$ or one half.

Circle the shape that shows halves.

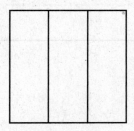

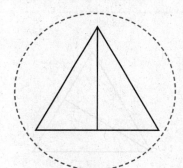

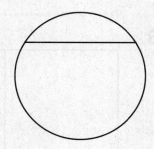

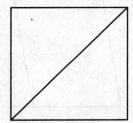

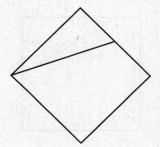

Circle the shape that shows halves.

1.

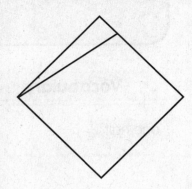 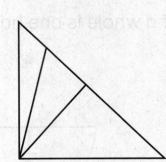 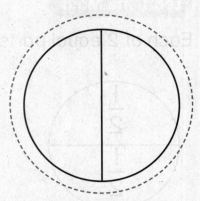

2.

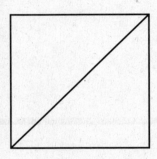 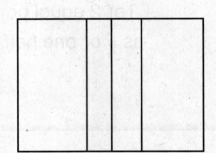 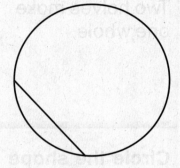

3.

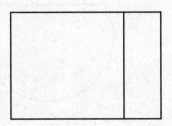 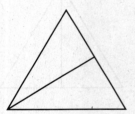 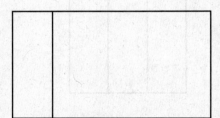

4.

 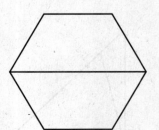 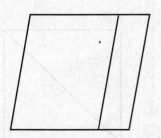

Name_____

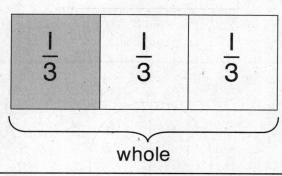

Learn the Math

A shape with thirds has 3 equal parts.

A shape with fourths has 4 equal parts.

1 of 3 equal parts is $\frac{1}{3}$ or one third.

Vocabulary

one third $\frac{1}{3}$

one fourth $\frac{1}{4}$

1 of 4 equal parts is $\frac{1}{4}$ or one fourth.

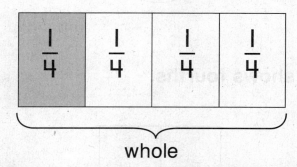

Circle the shape that shows thirds.

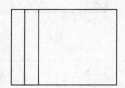  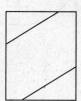

Circle the shape that shows fourths.

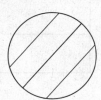  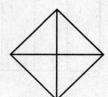

© Houghton Mifflin Harcourt

Circle the shape that shows thirds.

1.

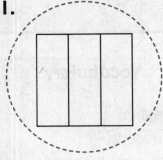 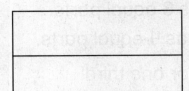 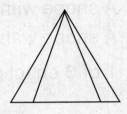

2.

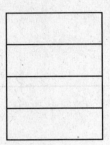 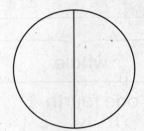 

Circle the shape that shows fourths.

3.

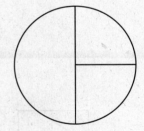 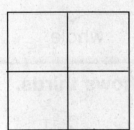 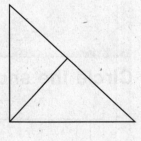

4.

 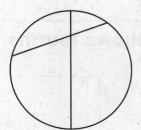 

Learn the Math

You can use models to show numbers to 50.

Tens	Ones

34

_3_ tens _4_ ones = _34_

Write how many tens and ones.
Write the number.

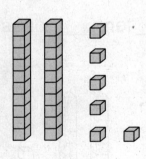

_2_ tens _6_ ones = _26_

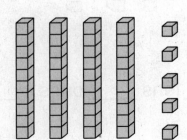

____ tens ____ ones = ____

Do the Math

Use ▭▭▭▭▭▭▭▭ and ▫.
Write how many tens and ones. Write the number.

1.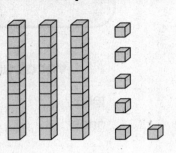

___3___ tens ___6___ ones = ___36___

2.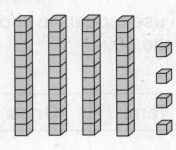

_____ tens _____ ones = _____

3.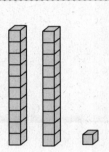

_____ tens _____ one = _____

4.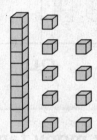

_____ ten _____ ones = _____

5.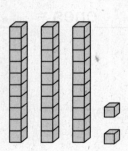

_____ tens _____ ones = _____

6.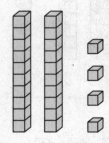

_____ tens _____ ones = _____

Name_____

© Houghton Mifflin Harcourt

Learn the Math

You can use models to show two-digit numbers to 100.

Tens	Ones

62

6 tens 2 ones

62

Write how many tens and ones.
Write the number.

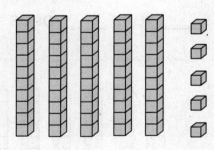

5 tens 5 ones

55

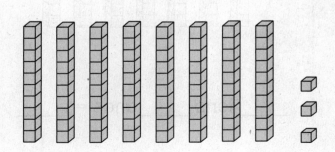

8 tens 3 ones

83

Use 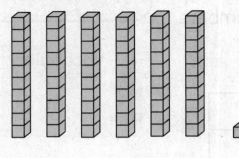 and ☐.
Write how many tens and ones. Write the number.

1.

_6_ tens _1_ one = _61_

2.

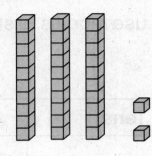

____ tens ____ ones = ____

3.

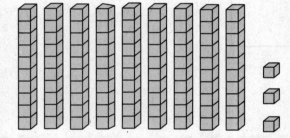

____ tens ____ ones = ____

4.

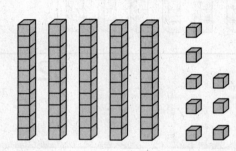

____ tens ____ ones = ____

5.

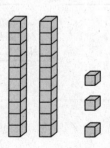

____ tens ____ ones = ____

6.

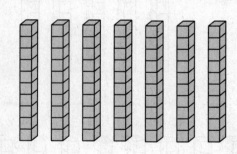

____ tens ____ ones = ____

Learn the Math

You can use models to compare numbers.

Write the numbers. Circle the greater number.

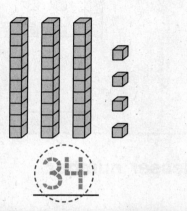 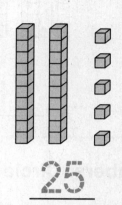

34   25

Write the numbers. Circle the lesser number.

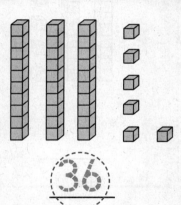 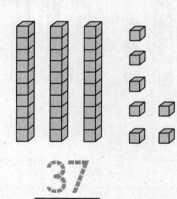

36   37

Write the numbers. Circle the numbers that are equal.

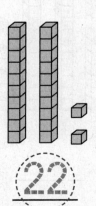 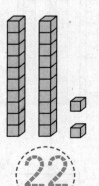 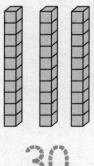

22   22   30

Write the numbers. Circle the greater number

1.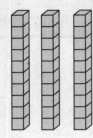

30   (31)

2.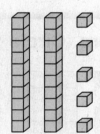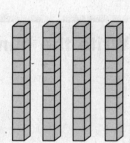

_____   _____

Write the numbers. Circle the lesser number.

3.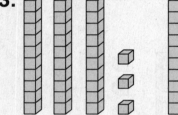

_____   _____

4.

_____   _____

Write the numbers. Circle the numbers that are equal.

5.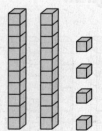

_____   _____   _____

Name_____

Learn the Math

You can use a number line to find the number that is before, between, or after other numbers.

Vocabulary

after
before
between

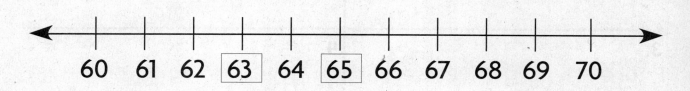

60 61 62 [63] 64 [65] 66 67 68 69 70

63 is just before 64.

64 is between 63 and 65.

65 is just after 64.

Write the number that is just before, between, or just after.

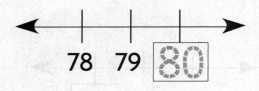

78 79 [80]

80 is just after 79.

© Houghton Mifflin Harcourt

Write the number that is just before, between, or just after.

1.

51 52 53

2.

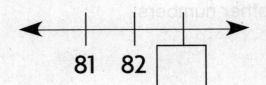

81 82 ☐

3.

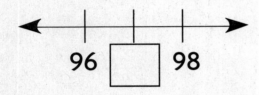

96 ☐ 98

4.

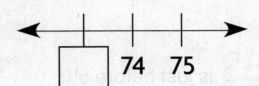

☐ 74 75

5.

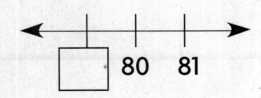

☐ 80 81

6.

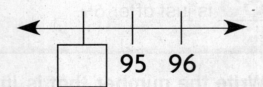

☐ 95 96

7.

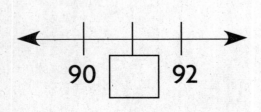

90 ☐ 92

8.

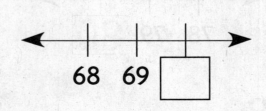

68 69 ☐

Learn the Math

You can use ▱▱▱ to add.

Add 26 and 5.

Step 1 Show 26 and 5.

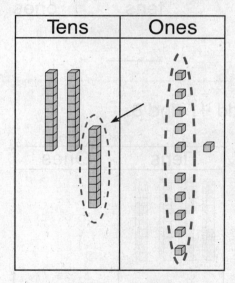

Tens	Ones

Step 2 Can you make a ten?
If you can, regroup
10 ones for 1 ten.

Tens	Ones

Step 3 Write how many tens
and ones. Write the sum.

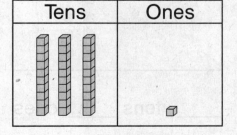

Tens	Ones

<u>3</u> tens <u>1</u> one

_____

Use ▭ ▫ to add. Write how many tens and ones.
Write the sum.

1. Add 32 and 4.

Tens	Ones

___3___ tens ___6___ ones

_____

2. Add 27 and 6.

Tens	Ones

_____ tens _____ ones

3. Add 41 and 8.

Tens	Ones

_____ tens _____ ones

_____

4. Add 34 and 7.

Tens	Ones

_____ tens _____ one

_____

Learn the Math

You can use mental math to add tens.

50 + 20 = _?_

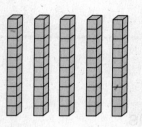 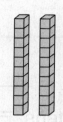

> **Think:** 5 + 2 = 7
>
> So, 5 tens + 2 tens = __7__ tens.

7 tens is equal to __70__.

So, 50 + 20 = __70__ .

─────────────────────────────────

20 + 30 = _?_

> **Think:** 2 + 3 = 5
>
> So, 2 tens + 3 tens = __5__ tens.

5 tens is equal to _____.

So, 20 + 30 = _____ .

Add.

1. $4 + 3 = \underline{7}$

   $\underline{4}$ tens $+ \underline{3}$ tens $= \underline{7}$ tens

   $40 + 30 = \underline{70}$

2. $5 + 4 = \underline{}$

   5 tens $+ 4$ tens $= \underline{}$ tens

   $50 + 40 = \underline{}$

3. $2 + 6 = \underline{}$

   2 tens $+ 6$ tens $= \underline{}$ tens

   $20 + 60 = \underline{}$

4. $\begin{array}{r} 30 \\ + 60 \\ \hline \end{array}$	5. $\begin{array}{r} 10 \\ + 10 \\ \hline \end{array}$	6. $\begin{array}{r} 20 \\ + 70 \\ \hline \end{array}$
7. $\begin{array}{r} 40 \\ + 40 \\ \hline \end{array}$	8. $\begin{array}{r} 60 \\ + 10 \\ \hline \end{array}$	9. $\begin{array}{r} 10 \\ + 20 \\ \hline \end{array}$

Name_____

Learn the Math

You can use ▱▱▱▱ ▱ to subtract.

Subtract 8 from 36.

Step 1 Show 36.

Tens	Ones

Step 2 Can you subtract 8
ones? If not, regroup
10 ones as 1 ten.

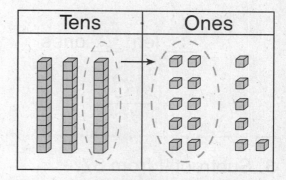

Step 3 Subtract 8 ones.

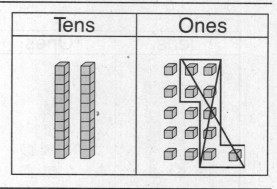

Step 4 Write how many tens
and ones.

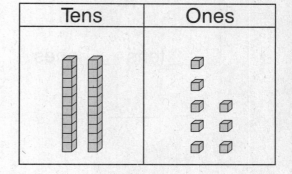

___ tens ___ ones

So, 8 from 36 is _____ .

© Houghton Mifflin Harcourt

Use ▱ ▱ to subtract. Write how
many tens and ones. Write the difference.

1. Subtract 5 from 21.

Tens	Ones

__1__ ten __6__ ones

__16__

2. Subtract 3 from 46.

Tens	Ones

____ tens ____ ones

____

3. Subtract 4 from 37.

Tens	Ones

____ tens ____ ones

____

4. Subtract 7 from 34.

Tens	Ones

____ tens ____ ones

____

Learn the Math

You can use mental math to subtract tens.

60 − 40 = _?_

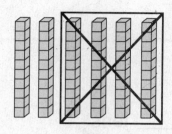

> **Think:** 6 − 4 = 2
>
> So, 6 tens − 4 tens = _2_ tens.

2 tens is equal to _20_.

So, **60 − 40** = _20_.

────────────────────────────────

40 − 10 = _?_

> **Think:** 4 − 1 = 3
>
> So, 4 tens − 1 ten = _3_ tens.

3 tens is equal to _____ .

So, **40 − 10** = _____ .

Subtract.

1. $7 - 5 =$ __2__

7 tens $- 5$ tens $=$ __2__ tens

$70 - 50 =$ __20__

---

2. $8 - 4 =$ ___

8 tens $- 4$ tens $=$ ___ tens

$80 - 40 =$ ___

---

3. $5 - 3 =$ ___

5 tens $- 3$ tens $=$ ___ tens

$50 - 30 =$ ___

4.
$$\begin{array}{r} 90 \\ -\ 50 \\ \hline \end{array}$$

5.
$$\begin{array}{r} 70 \\ -\ 30 \\ \hline \end{array}$$

6.
$$\begin{array}{r} 40 \\ -\ 30 \\ \hline \end{array}$$

7.
$$\begin{array}{r} 50 \\ -\ 10 \\ \hline \end{array}$$

8.
$$\begin{array}{r} 80 \\ -\ 50 \\ \hline \end{array}$$

9.
$$\begin{array}{r} 90 \\ -\ 70 \\ \hline \end{array}$$

Name_____

Learn the Math

You can skip-count by twos and fives.

Skip-count. Count the mittens by twos.

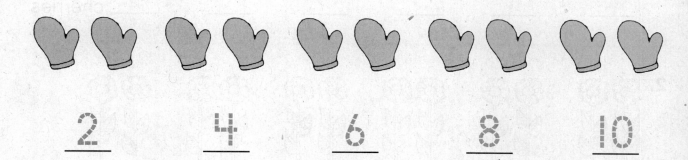

$$\underline{2} \qquad \underline{4} \qquad \underline{6} \qquad \underline{8} \qquad \underline{10}$$

So, there are ___10___ mittens in all.

Skip-count. Count the fingers by fives.

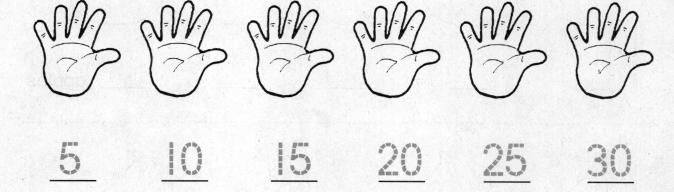

$$\underline{5} \qquad \underline{10} \qquad \underline{15} \qquad \underline{20} \qquad \underline{25} \qquad \underline{30}$$

So, there are _____ fingers in all.

Do the Math

Skip-count by twos or fives. Write how many.

1.

2   4   6   8   10   12 cherries

2.

_____ _____ _____ _____ _____ _____ flowers

3.

_____ _____ _____ _____ _____ apples

4.

_____ _____ _____ _____ bananas

Name_____

Learn the Math

You can use a hundred chart to skip-count.

1	2	3	4	(5)	6	7	8	9	(10)
11	12	13	14	(15)	16	17	18	19	(20)
21	22	23	24	(25)	26	27	28	29	(30)
31	32	33	34	(35)	36	37	38	39	(40)
41	42	43	44	(45)	46	47	48	49	(50)
51	52	53	54	(55)	56	57	58	59	(60)
61	62	63	64	(65)	66	67	68	69	(70)
71	72	73	74	(75)	76	77	78	79	(80)
81	82	83	84	(85)	86	87	88	89	(90)
91	92	93	94	(95)	96	97	98	99	(100)

Use the hundred chart to skip-count.
Start at 24. Skip-count by twos.

24, _26_, _28_, _30_, _32_, _34_, _36_

Start at 20. Skip-count by fives.

20, _25_, _30_, _35_, _40_, _45_, _50_

Do the Math

Use the hundred chart to skip-count.

1	2	3	4	5	6	7	8	9	10
11	12	13	14	15	16	17	18	19	20
21	22	23	24	25	26	27	28	29	30
31	32	33	34	35	36	37	38	39	40
41	42	43	44	45	46	47	48	49	50
51	52	53	54	55	56	57	58	59	60
61	62	63	64	65	66	67	68	69	70
71	72	73	74	75	76	77	78	79	80
81	82	83	84	85	86	87	88	89	90
91	92	93	94	95	96	97	98	99	100

I. Start at 15. Skip-count by fives.

15, 20, 25, 30, 35, 40, 45

2. Start at 42. Skip-count by twos.

42, ____, ____, ____, ____, ____, ____

3. Start at 30. Skip-count by fives.

30, ____, ____, ____, ____, ____, ____

4. Start at 20. Skip-count by tens.

20, ____, ____, ____, ____, ____, ____